Dedicated to the One Beyond the Universe

Written by Brandon Nguyen and Illustrated by Melissa Lam.
First Edition, 2026
ISBN: 979-8-950362-02-6
Printed in USA

ZOOM TO THE PLANETS

Written by Brandon Nguyen
Illustrated by Melissa Lam

Hop in my rocket, it's shiny and bright.
We'll zoom through the stars on a magical night.
With a **WHOOSH** and a **ZOOM,** we'll be off on our way.
To visit the planets and giggle and play!

COUNT DOWN WITH ME!
5 4 3 2 1 ... BLAST OFF!

First stop is Mercury-speedy and small.
Racing round the Sun, the fastest of all.
It's **ROCKY** and **BOILING,** a hot, fiery place.
We'll wear silver sun-hats in this sizzling space!

PRETEND YOUR FEET ARE BURNING

HOP, HOP, HOP!

Venus is next, it's shiny and bright.
Glowing like lanterns that glow in the night.
But oh! It's so steamy, so **MISTY** and **HOT**.
We'll fan ourselves quickly - this heat's quite a lot!

WAVE YOUR HANDS

LIKE YOU'RE FANNING AWAY STEAM!

Here is our Earth, so **LOVELY** and **ROUND**.
With oceans and forests and life all around.
Our home full of laughter, of rivers and song.
It's here we belong - Yes, here we belong!

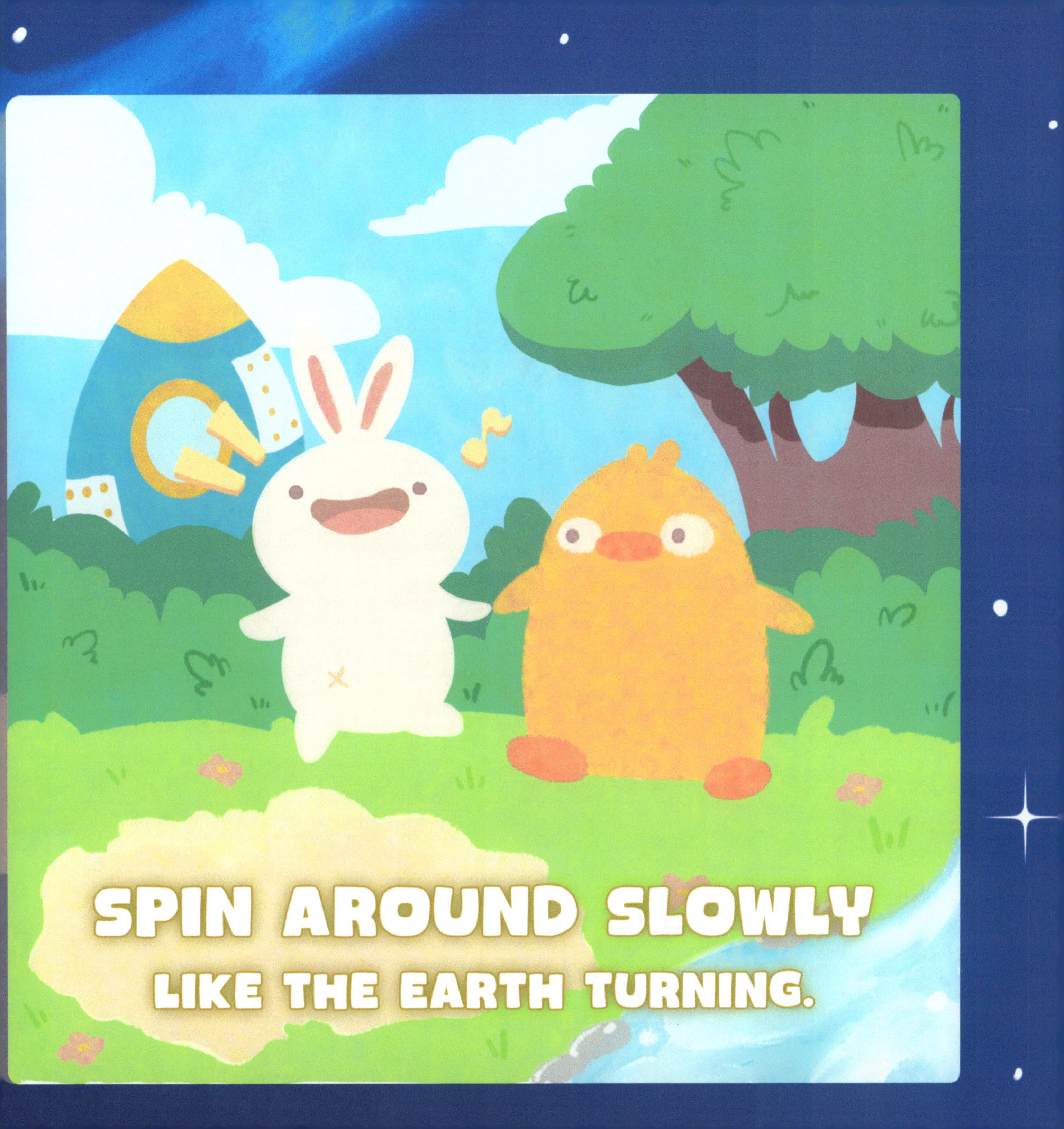

SPIN AROUND SLOWLY
LIKE THE EARTH TURNING.

Onward to Mars, the planet's so red,
with rust-colored rocks where robots now tread.
Someday, perhaps, we'll live there and play,
in little red houses - imagine that day!

MARCH LIKE A ROBOT

CLUNK, CLANK, CLONK!

Look at big Jupiter - **SWIRLY** and **GRAND**.
The largest of planets in all of space - land.
Its Great Red Spot is a storm that won’t end.
Let’s swirl round and round in a stormy pretend!

WAVE YOUR ARMS
LIKE A GIANT WINDY STORM.

Now off to Saturn with rings shining bright.
They **GLIMMER** and **SHIMMER**, a beautiful sight!
We'll slide on the ice-rings, we'll twirl, and we'll sing.
A whizzing, ring-dingy, round-the-planet fling!

SPIN IN A CIRCLE
LIKE SATURN'S RINGS.

Uranus is funny - it tilts on its side.
It rolls as it spins, like a ball on a ride.
It's **CHILLY** and **ICY**, you'll glide in your shoes!
Let's roll on our sides like Uranus, too!

LIE ON YOUR SIDE AND
ROLL GENTLY!

Neptune is last, so stormy and blue.
Its winds are the fastest, they'll whoosh right past you!
It's **WILD** and it's **GUSTY**, a far-away show.
The windiest place you could ever go!

TAKE A DEEP BREATH

BLOW LIKE NEPTUNE'S WIND!

We've zipped through the planets, from the first to the last.
And now we're back home, that trip was a blast!
For Earth is our home, full of love, life, and cheer.
The most magical planet is right...right here!

GIVE YOURSELF A HUG

EARTH IS HOME!

EARTH IS OUR HOME!

www.ingramcontent.com/pod-product-compliance
Lightning Source LLC
LaVergne TN
LVHW071238210926
848486LV00007B/299